I love to watch freight trains with my Uncle Ethan. Now I hear the train whistle blowing outside. Then comes the rumble of wheels. "Whooo-whooo! Train on the way!" I proclaim.

Uncle Ethan grabs his cap. His quick reply is, "Hurry, let's go!"

Train tracks run near Uncle Ethan's house. But you cannot see much through the dark green thicket hedge. So we scurry across the yard and up the hill.

From there, we can watch whole trains wind their way across the valley. A whistle will often pierce the air. The freight trains roll along hauling cargo, such as food, lumber, or cattle.

On the hilltop, Uncle Ethan and I sit together and count cars. One, two, five, eight, eighteen ... twenty-eight ... thirty-eight ... forty-eight. Clickety-clack, clickety-clack!

In between, we take turns naming the freight cars. I love the sound of their names—flatcar, boxcar, hopper, caboose. Uncle Ethan loves trains almost as much as I do! Or maybe even more!

"Could we go on a train together?" I ask. "I'll buy our tickets!" I declare. "We could zigzag zoom along!"

Uncle Ethan nods, smiling a secret kind of smile. "Zigzag zoom along," he says as an echo.

Then, the fantastic train tales start.

"See that train there?" I ask Uncle Ethan. "Let me describe to you what is inside. Why, this tale is sure to astound you! The train is filled with eagles, egrets, and cuckoos. They are headed for a Russian wildlife park far across the sea!"

"Is that so?" Uncle Ethan responds. "How will the train ever be able to get across the sea?"

"A freight tanker will ferry it over," I answer.

Then Uncle Ethan adds to the tale. He says, "I myself thought that train was filled with zebras. Eight hundred zebras!"

But before his tale is done the train is gone. Our train patrol has ended for the moment.

Down the hill we track again, the zebra tale left untold. Uncle Ethan and I, walking hand in hand, are silent now.

But once home, I pose a crucial question: "Where were the eight hundred zebras going, Uncle Ethan?"

Uncle Ethan smiles his secret smile. "Just zigzag zooming along," he answers. "Zigzag zooming along *together*."